# POLAR BEARS

## Jolyon Goddard

**Grolier**
an imprint of
**■SCHOLASTIC**
www.scholastic.com/librarypublishing

Published 2008 by Grolier
An imprint of Scholastic Library Publishing
Old Sherman Turnpike, Danbury,
Connecticut 06816

**For The Brown Reference Group plc**
Project Editor: Jolyon Goddard
Copy-editors: Ann Baggaley, Lisa Hughes
Picture Researcher: Clare Newman
Designers: Jeni Child, Lynne Ross,
    Sarah Williams
Managing Editor: Bridget Giles

Volume ISBN-13: 978-0-7172-6277-9
Volume ISBN-10: 0-7172-6277-4

**Library of Congress
Cataloging-in-Publication Data**

Nature's children. Set 3.
    p. cm.
    Includes bibliographical references and
index.
    ISBN 13: 978-0-7172-8082-7
    ISBN 10: 0-7172-8082-9
    1. Animals--Encyclopedias, Juvenile. I.
    Grolier Educational (Firm)
    QL49.N384 2008
    590.3--dc22
                            2007031568

    445 1474

Printed and bound in China

# Contents

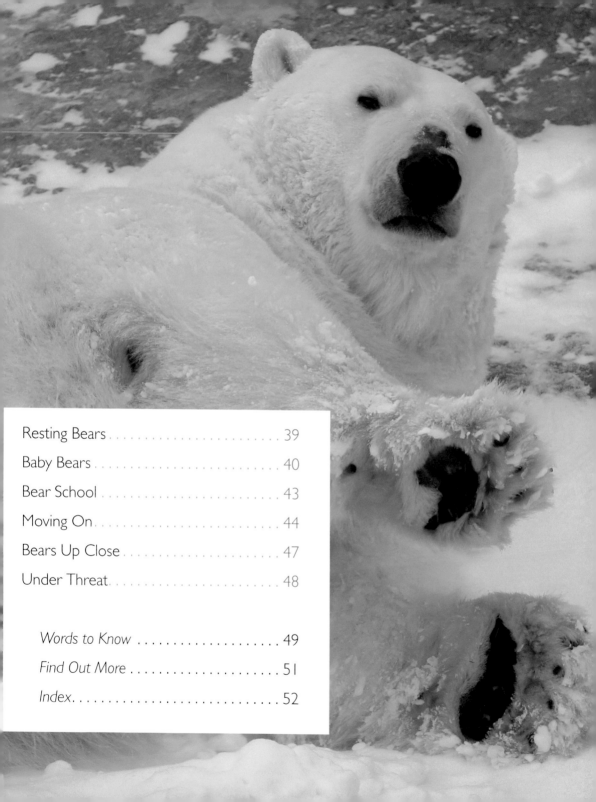

# FACT FILE: Polar Bears

| | |
|---|---|
| **Class** | Mammals (Mammalia) |
| **Order** | Carnivores (Carnivora) |
| **Family** | Bear family (Ursidae) |
| **Genus** | Black, brown, and polar bears (*Ursus*) |
| **Species** | Polar bear (*Ursus maritimus*) |
| **World distribution** | Ice covering the Arctic Ocean and the northern lands that border it: Alaska (United States), Canada, Greenland (Denmark), Svalbard (Norway), and Russia |
| **Habitat** | Sea ice and coastal areas; some polar bears move inland during summer |
| **Distinctive physical characteristics** | White, creamy fur; massive body, with long neck, and small head and ears; short tail; big paws with tufts of fur between the toes |
| **Habits** | Live alone except females raising young and during the mating season; polar bears are excellent swimmers; they often travel very great distances to find food |
| **Diet** | Mostly sea animals, especially seals; also eat small whales, young walrus, and fish; in late summer, grasses, lichens, and berries |

# Introduction

For many people, the polar bear, also called the ice bear, ice king, or *nanook*, symbolizes the wilderness of the Arctic. This magnificent hunter is without doubt the king of the Arctic. The largest land **predator**, the polar bear is an expert hunter of seals and other animals. It easily tracks down these animals using its amazing sense of smell. One swipe of a polar bear's powerful paw knocks out or kills most of its **prey**. Though mostly a land animal, the polar bear is also at home in the freezing Arctic waters, swimming great distances in search of seals.

**A young polar bear shelters under its mother.**

Some brown bears are as big as polar bears.

6

# Bear Family

The bear family is small. There are just eight types, or species, of bears. Three of these species can be found in North America. They are the polar bear, brown bear, and American black bear.

Grizzly bears are brown bears with frosted, or "grizzled," fur. They live solely in North America. Brown bears also live in Europe and Asia.

The Andean bear is the only bear that lives in South America. The mixture of light and dark fur on this bear's face sometimes makes it look like it is wearing glasses, which is why the Andean bear is also called the spectacled bear. (Spectacles are another name for glasses.) The four other bears are sloth bears, Asian black bears, sun bears, and giant pandas. They all live in Asia.

Bears are **carnivores**, or meat eaters. Those are mammals that have sharp teeth **adapted** to eating meat. Other carnivores include cats, dogs, and raccoons. Most bears, however, eat both plants and animals. Only the giant panda eats mostly plants—mainly the grass bamboo—whereas polar bears are almost exclusively meat eaters.

**7**

# Polar Bear History

Polar bears **evolved** rapidly from an isolated group of brown bears—with a taste for seals— about 200,000 years ago. The bears quickly adapted to their surroundings. Their fur became white to help them stay **camouflaged** against the ice and snow. Their teeth became sharper, better for a diet of animals. Other bears have flatter teeth suited to grinding up plants. The bear's body shape changed, too—it became narrower and streamlined—to better suit a life that involved swimming. Their senses of smell and sight improved to give them a better chance of survival in the Arctic. This new type of bear soon spread across the Arctic, following its favorite food: seals.

Polar bears can still breed with brown bears. That rarely happens in the wild as their **habitats** do not usually overlap. However, it occasionally happens in zoos. Scientists have shown that the brown bears living on the ABC Islands off Alaska are more closely related to polar bears than they are to other brown bears! These bears no doubt share a common ancestor with polar bears.

Polar bears are the most recent type of bear species to evolve.

Beneath the fur, the polar bear's skin is black, as can be seen on the bare nose.

# Bear Essentials

Some male polar bears, or **boars**, are huge. They weigh as much as 1,700 pounds (770 kg). That's the same weight as a small car. These males grow up to 8⅓ feet (2.5 m) long from head to tail. When a full-grown male stands upright on his back legs he might be 12 feet (3.7 m) tall. Females, or **sows**, weigh much less, up to 750 pounds (340 kg) and measure up to 6¼ feet (1.9 m) long.

The polar bear has a different body shape from its ancestor the brown bear. It has a more wedge-shaped body, narrower at the front, with back legs that are longer than the front ones. The polar bear's head is smaller than a brown bear's and the Arctic species has a slightly arched snout in profile, unlike the "dished," or concave, profile of the brown bear. The polar bear's neck is longer, too. Like other bears, the polar bear has a short tail. Its eyes are small and so are its ears. If the ears were larger, they would lose a lot of heat and get damaged by the freezing Arctic air.

# Arctic Home

Polar bears live in the far north, a region called the Arctic. Few animals are tough enough to survive in the extreme cold and fierce winds of this treeless region. In winter, temperatures drop to -50°F (-40°C). In summer there is constant daylight.

Polar bears live where they can find seals. That might be on the edge of the ice that covers the Arctic Ocean or on the coasts of lands and islands surrounding the ocean. These regions are part of just five countries: United States (Alaska), Canada, Denmark (Greenland), Norway (Svalbard), and Russia.

The polar bears have no fixed **territory**, unlike other bears. As the ice expands in winter and melts in summer, they might roam in areas as large as 115,000 square miles (300,000 sq km), searching for seals.

In summer, some polar bears move inland. There, they eat berries, plants, and small animals. Other bears follow the seals in summer as they move north, searching for fish.

Most polar bears live near the edge of the Arctic ice because that's where seals are most likely to be.

13

**Polar bears keep their fur in tiptop condition. They spend a lot of time grooming and will wash off blood and grease after each meal.**

# Fabulous Fur

One of the secrets of a polar's success is its fur. The bear's coat has two layers. An outer layer of long hairs called **guard hairs** and an inner layer of shorter woolly hairs called **underfur**. The shiny, hollow guard hairs act like a raincoat. They repel water, ice, and snow. The underfur traps air that is warmed by the bear's body heat. It's like a cozy blanket surrounding the bear.

Close up the fur isn't actually white—it's see-through. It appears ivory white or creamy because it reflects sunlight. In older bears, the fur often appears to be a yellowy gold. The see-through hairs act a little like the glass of a greenhouse, warming the bear's skin.

Against the white snow and ice, the bear blends in with its surroundings. This camouflage allows the bear to sneak up unseen on basking seals. Polar bears that move inland during summer stand out against their snowless background. But that is not a problem because during this time the bears do not hunt seals, and no other animals are likely to attack the bears.

# Powerful Paws

A polar bear's huge paws can measure up to 13 inches (33 cm) long and 9 inches (23 cm) wide. The hind paws are somewhat bigger than the forepaws. Each paw has five toes, and each toe has a claw. The claws are about 2 inches (5 cm) long. There are tufts of thick hair between the toes on each paw. That not only keeps them warm but also helps them grip on the snow and ice when the bear is walking. The bear uses its paws to kill its prey. One swipe of a forepaw is usually enough to kill or knock out most prey.

The big paws benefit the bear in another way: they act as snowshoes, spreading out the polar bear's weight and stopping it from sinking too deep into snow. On weak ice, the bear crawls along on the surface, spreading its weight farther to stop the ice from breaking.

Unlike a pet cat,
a polar bear cannot
retract, or pull in,
the black claws on
the end of each toe.

17

Polar bears walk slowly.
They are at risk of
overheating if they
move too quickly.

# Pigeon-toed Walk

Polar bears walk on the flats of their feet—just like humans do. They are also pigeon-toed. That means that their feet point slightly inward as they walk. That might look clumsy, but it suits a polar bear just fine. Polar bears do a lot of walking. They might cover 20 miles (32 km) in a single day. They walk along at about 2.5 miles (4 km) per hour. When they need to, they can run fast—up to 25 miles (40 km) per hour—but just for short distances.

Where polar bears live the land or ice is usually flat. But sometimes hills of snow and ice form. The bears can climb these inclines with ease. Coming down, the bears either walk cautiously, using their front legs as brakes, or they slide down on their back, belly, or side headfirst, backward—looking over their shoulder—or even sideways!

# Strong Swimmers

The scientific name for the polar bear *Ursus maritimus* means "bear of the sea." The polar bear lives up to its name: it is an excellent swimmer. In water, the bear dog-paddles, using its front legs to do all the swimming work and the back legs to steer. The bear has a streamlined body that is narrower at the front. It cuts through water with ease. Polar bears do not swim fast, but they can swim for very long distances, up to 60 miles (100 km) without a break.

The polar bear's long neck keeps its eyes and nostrils above the surface of the sea. Sometimes, the bear dives underwater to avoid a floating chunk of ice or to chase a seal. As it does that, it closes its nostrils tightly to prevent water from entering its air passage. Polar bears can hold their breath for about 2 minutes.

When the bear comes out of the water, it shakes itself dry, just like a pet dog does. It might even roll in the snow, which removes excess water, like a towel.

Some polar bears journey up to 2,000 miles (3,200 km) across the ocean, combining long swims with hitching rides on floating ice.

A polar bear's favorite food is seal blubber, which can quickly be processed and added to the bear's own protective blubber.

# Blubbery Bears

The fur may keep a polar bear warm and waterproof on land, but it is not that effective at keeping the bear warm when it is swimming in the bitterly cold sea. However, the bear is still protected from the cold waters. It has a very thick layer of fat under the skin. This fat is called **blubber**. Many other marine mammals, such as whales, seals, and walrus, have a thick layer of blubber, too.

A polar bear's blubber might be up to 3.5 inches (9 cm) thick around some parts of its body. The blubber insulates, or stops body heat from escaping, as the bear swims. The blubber also acts as an energy store. When the bear has to go without food—as many polar bears do, often for several months when prey is hard to find—the bear slowly breaks down the blubber to fuel its body.

# Super Senses

Finding enough to eat in the Arctic can be very difficult. Polar bears have amazing senses to help them find food. Even though their eyes are small, polar bears have sharp eyesight. That allows them to see prey, such as a basking seal, from a distance. Polar bears often stand on their back legs to get a better view of their surroundings. Their hearing is good, too.

However, the polar bear's sharpest sense is smell. That is just as well because polar bears usually hunt in the dark, the time when seals are most active. On the Arctic ice, a polar bear can pick up the smell of a seal from 0.6 miles (1 km) away, even if the seal is hidden in its **den** 40 inches (1 m) below the snow.

# Hungry Hunters

Polar bears spend a lot of time hunting. Seals are their favorite food, especially young seals, which are 50 percent blubber! But they can be tricky to catch. In winter seals hunt fish under the ice. But the seals have to come up for air every so often. They do that at breathing holes, which the seal makes by scraping away the ice roof of the sea with its flippers.

If a polar bear finds a breathing hole, it will lie beside it, waiting patiently for a seal to poke its head out. When the seal does that, the bear attempts to grab the seal with its paws and jaws. An adult bear usually just eats the seal's blubber. The rest of the seal is left for scavenging animals, such as Arctic foxes, gulls, and younger, less experienced polar bears. If the seal gets away, the bear—often after an angry outburst— will go in search of another breathing hole.

A sow travels with her three young, or cubs. When they get tired, they ride on her back.

# Stalking Seals

Although a polar bear prefers to lie and wait at a seal's breathing hole, it will also stalk, or creep up on, seals. The bear follows its nose until it spots a seal, resting on ice. The bear creeps slowly toward the seal, its fur keeping it camouflaged and the tufts of hair on its paws muffling its footfalls. When the bear is in striking distance, it pounces like a cat onto the unsuspecting seal. A polar bear might also catch a seal by swimming up to it as it rests on a floating piece of ice. The bear then leaps out of the water onto the seal.

The ringed seal is the bears' favorite prey, but they also like bearded and hooded seals. The bears will also eat young walrus. Sometimes a large group of polar bears will gather around and share a washed up adult walrus or whale.

In winter groups of beluga and narwhals often get trapped in pockets of water surrounded by ice. Bears gather around the opening and try to catch one of the small whales each time it surfaces for air. The bears often stay there for weeks until they've caught and eaten every trapped whale.

Polar bears find most of their food between April and mid-July, when there are plenty of seal pups.

Fewer than 10 people in Alaska and Canada have been killed by polar bears in the past 30 years.

# Few Enemies

Very few animals will fight with a polar bear. When food is particularly hard to find, a polar bear might attack an adult walrus. The walrus will defend itself with its long, sharp tusks. Those can seriously wound or kill a polar bear.

Other than humans, a polar bear's main enemy is other polar bears. Usually, however, bears will go out of their way to avoid one another. But in the **mating season**, boars will fight—sometimes to death—over a sow. Males will also kill young bears, or **cubs**, if the mother doesn't defend them adequately. Some people think that orcas, or killer whales, attack polar bears in the sea, but as of yet there is no evidence of that happening.

# Life on Land

In summer, some polar bears go north, traveling hundreds of miles, as the Arctic ice retreats. Others come ashore and wander inland. A polar bear's fur and blubber are so effective at keeping it warm that in summer the bear often gets too hot. It sheds, or **molts**, some fur to lose heat.

At this time of the year the bear has to take it easy. Any vigorous exercise will make it overheat. Bears lose heat through their paws. They can also pant, like a dog. Some bears dig down to where the ground is still frozen and rest, cooling off.

On land, males may form pairs, often play-fighting. They are not aggressive toward each other because there is little food and no sows to fight over. The food that is available includes berries, roots, birds, and small mammals.

When there's no food the bears slow down, entering a state called "walking hibernation" to save energy. No other mammal is known to do that. When food becomes available, the bears come out of this state. By fall, the sea begins to freeze over and the bears return to the sea ice.

A young bear relaxes and cools itself with a roll in the ice.

33

From one polar bear skin, Inuits could make two pairs of trousers and a pair of boots called kamiks.

# Inuits and Bears

Inuits—natives of Arctic Canada—called the polar bear *nanook*. In their legends, polar bears were very similar to humans. They walked on two legs and could talk to one another. In the privacy and comfort of their own homes, the bears would take off their fur coats.

Inuits hunted polar bears. Sometimes the bears hunted the Inuits! Inuits greatly respected the polar bear. A male Inuit became a man after he had killed his first polar bear. A hunter would offer any polar bear he killed tools to use in its afterlife. By honoring a bear in such a way, it would communicate with living bears, telling them to come and be killed by the hunter. A bear not properly honored would warn other bears to avoid the disrespectful hunter.

Inuits ate polar bear meat. However, they were careful to avoid the bear's liver, which stores vitamin A. This vitamin is poisonous to humans in such high levels as those found in a polar bear's liver. Inuits also made clothes from polar bear skins.

# Mating and Denning

Sows are ready to have cubs at four or five years old. Boars approach sows from late March to early June. The sow will **mate** with the biggest and strongest boar. They stay together from anywhere from a few days to a few weeks. They then go their separate, solitary ways. The boar will have no role in looking after his young.

In October, the pregnant sow looks for a place to have her cubs. She looks for a south-facing snowy hillside. The sow digs out a room about 40 inches (1 m) high in the snowdrift. The entrance tunnel is between 5 and 10 feet (1.5–3 m) long and lower than the room. That allows any water from melted snow to run out of the den and warm air, which rises, to be trapped in the den. Once the den is completed, the sow curls up inside and sleeps, while her babies develop inside her. Her breathing and heart rate slow and her body temperature drops, but she is not truly hibernating. She could quickly come out of her sleepy state if she needed to.

Boars fight over a sow.
The fights can be fierce.
If one boar kills the
other, the winner
might eat the loser.

Polar bears like a lot of sleep. That helps them conserve energy when food is hard to find.

38

# Resting Bears

In winter, while pregnant sows are in their dens, other polar bears search for food. If the weather becomes particularly stormy a bear will dig a den in the snow and shelter there until the bad weather passes. A bear will never enter another bear's den, however.

Polar bears like to sleep much the same as humans—for 7 to 8 hours at a time. When a polar bear is tired it finds a sheltered spot or digs a shallow pit and sleeps outdoors. The bear sleeps curled up with its head resting in its chest, its front legs hugging the back legs, and its back to the brutal wind. The bear's breath warms the front of its body. Polar bears also like to have a snooze after they have eaten.

# Baby Bears

Baby polar bears, or cubs, are born between November and early January. The sow usually has two cubs, but sometimes she has just one or occasionally three or four.

Newborn cubs cannot hear, see, or smell, and they are covered in fine hair. They are tiny compared to their huge mother—about the size of adult guinea pigs, weighing 1⅓ pounds (600 g). They **nurse** on their mother's rich milk and grow quickly. By four weeks, they can hear, and a week later their eyes open. A couple of weeks after that, they can smell and walk.

By three months, the cubs have grown white, woolly fur and weigh about 22 pounds (10 kg). They are now ready to leave the cozy den and explore outside. The sow is very hungry. She hasn't eaten for about seven months. She has lived off her blubber. Outside the den, she will look under the snow for grass and lichens to eat. The den usually faces the sun. The sow likes to sunbathe outside, while the cubs nurse, lying on her belly.

When a sow leaves her den with her cubs in April she is about half the weight she was when she first entered the den in late fall.

**41**

Inuits call polar bear cubs *ah tik tok*, which means "those that go down to the sea."

# Bear School

The cubs learn their hunting skills by watching their mother. They pay attention as she sneaks up on seals or waits by a breathing hole. The cubs still nurse from her, but also begin to develop a taste for seals.

The sow is always aware of her cubs' needs. On their wanderings, she often stops to let the cubs nurse or have a rest on top of her. She licks them clean to keep their fur free of dirt. The cubs soon learn to groom themselves and one another.

The sow takes her cubs swimming. She dives in first. The nervous cubs whimper as their mother swims away, but soon follow. In no time at all, the cubs get used to the freezing water and start to enjoy themselves. If a cub gets tired during the swim, the sow lets it climb on her back.

# Moving On

One-year-old cubs are as big as a St. Bernard dog, weighing between 155 and 175 pounds (80–90 kg). Their mother still protects them from enemies that might attack a cub, such as wolves and aggressive polar bear boars. The cubs remain with their mother, learning from her, until they are about two and a half years old. That is the time when the sow is ready to mate again and raise another family.

The young bears then wander off on their own, rarely meeting other polar bears. A cub's first year alone is the toughest year of its life, and many die at this age. If the cub has been taught well by its mother it might live a long life—in the wild, sows can live to their early 30s and boars to their late 20s—and have several families of its own. In captivity, polar bears can live more than 40 years.

Play-fighting helps cubs develop skills that will help them survive as adults.

45

A tundra buggy approaches a sow and her cub. Tourists are forbidden to feed the bears.

# Bears Up Close

Polar bears are often one of the main attractions in zoos. However, the animals often don't look the same as in the wild. In zoos, tiny living things called cyanobacteria (SI-AN-NO-BAK-TEAR-E-UH) get inside the bears' hollow guard hairs. The cyanobacteria are bluey green and tinge the bears' fur with that color. Zookeepers sometimes bleach the bears' fur to whiten their coat. In addition, because the zoos are usually somewhere much warmer than the Arctic, the bears don't need a thick layer of blubber or a thick winter fur coat. Their fur might be so thin that their black skin shows through.

The best place to see polar bears is in the wild. Every year in fall polar bears wander close to and through Churchill, Manitoba, in Canada. This town is called "The Polar Bear Capital of the World." Tourists can take a ride on big buses called "tundra buggies" to get really close to the bears. The bears come up to the bus windows to sniff the humans. As far as the bears are concerned, the tourists would make a nice meal!

# Under Threat

There are about 25,000 polar bears in the wild. In the past, the polar bears' main threat was human hunters, who killed too many of the bears. In the 1970s Canada, Denmark, Norway, Russia, and United States agreed to protect the bears. However, hunters—mainly native peoples—are allowed to kill about 700 bears each year. A certain amount of illegal hunting also occurs.

These days, the polar bears' biggest threat is climate change. The planet has become slightly warmer, and much of the ice covering the North Pole has melted. Polar bears now have to travel farther to hunt seals. Scientists have been finding polar bears that have drowned after becoming exhausted from swimming great distances in search of seals. As the ice keeps melting, scientists think that the number of polar bears will decrease by one-third in the next 50 years.

In addition, pollution—such as from oil slicks and industrial waste—has affected polar bears. Poisonous chemicals affect the bears' ability to fight illness, making the bears prone to disease.

# Words to Know

**Adapted**        Changed to become better suited.

**Blubber**        A thick layer of fat under the skin.

**Boars**        Male polar bears.

**Camouflaged**    Having coloring or markings that help an animal blend in with its surroundings.

**Carnivores**    Meat-eating mammals with teeth suited to cutting and stabbing. Cats, dog, and bears are all carnivores.

**Cubs**        Young polar bears.

**Den**        An animal's home.

**Evolved**        Changed over many generations.

**Guard hairs**    Long, coarse hairs that make up the outer layer of a polar bear's coat.

**Habitats**        Places where animals or plants live.

**Mate**        To come together to produce young; either of a breeding pair of animals.

**Mating season**    The time of the year when animals come together to produce young.

**Molts**        Sheds fur and replaces it with new fur.

**Nurse**        To drink milk produced by the mother.

**Predator**        An animal that hunts other animals.

**Prey**        An animal hunted by other animals.

**Sows**        Female polar bears.

**Territory**        An area claimed by an animal that it defends against animals of the same type.

**Underfur**        A woolly layer of hair next to an animal's skin.

# Find Out More

**Books**

Osborne, M. P. and N. P. Boyce. *Polar Bears and the Arctic*.
Magic Tree House Research Guides. New York:
Random House Books for Young Readers, 2007.

Rosing, N. and E. Carney. *Face to Face with Polar Bears*.
Washington, DC: National Geographic Children's
Books, 2007.

**Web sites**

**Polar Bear**
*www.enchantedlearning.com/subjects/mammals/bear/
Polarbearcoloring.shtml*
Facts about polar bears and a picture to print and color in.

**Polar Bears: Creature Feature**
*www.nationalgeographic.com/kids/creature_feature/0004/
polar.html*
Information about polar bears, including a video.

# Index